T0194196

The Effects of Offshoring on the Information Technology Sector

Is it really affecting us?

MOHAMMED K. YUSUF

iUniverse, Inc.
Bloomington

The Effects of Offshoring on the Information Technology Sector
Is it Really Affecting Us?

iUniverse books may be ordered through booksellers or by contacting:

iUniverse
1663 Liberty Drive
Bloomington, IN 47403
www.iuniverse.com
1-800-Authors (1-800-288-4677)

Because of the dynamic nature of the Internet, any web addresses or links contained in this book may have changed since publication and may no longer be valid. The views expressed in this work are solely those of the author and do not necessarily reflect the views of the publisher, and the publisher hereby disclaims any responsibility for them.

Any people depicted in stock imagery provided by Thinkstock are models, and such images are being used for illustrative purposes only.

Certain stock imagery © Thinkstock.

ISBN: 978-1-4502-8892-7 (sc)
ISBN: 978-1-4502-8893-4 (ebook)

Printed in the United States of America

iUniverse rev. date: 1/18/2011

Acknowledgments

I would like to take this chance to thank my mother, Iyabo A. Yusuf. Without her love and support, this book would not have been possible.

Abstract

The information technology (IT) field has continued to change over the past few years. The objective of this thesis is to investigate the effects of offshoring on the information technology field through a study of reasons for offshoring. Utilizing several studies conducted over the past fourteen years, this paper compares and contrasts the types of jobs that have been offshored. Charts from different sectors of the IT industry and charts comparing jobs lost and created due to offshoring all have been included. Results show that offshoring is affecting the lower-level jobs in the IT field, and that higher-level business decision-making rolls are flourishing in this market. Information technology jobs have been and will continue to be sent offshore; however, this study shows that the high-level IT jobs are staying onshore and offshoring is encouraging low-level IT professionals to broaden their horizons.

Contents

Description of Problem

Statement of Problem

Data on employment trends by industry and occupation suggest that offshoring in the information technology (IT) sector occurs, but has historically not had significant impact (Bednarzik 2005). However, if current trends continue, within the next five years, 800,000 of the 10 million IT jobs in the United States will move offshore (Bednarz 2004. These forecasted trends, and the current and anticipated impacts of offshoring on the IT sector, are the focus of this research.

Purpose of Project

This research will investigate the effects of offshoring on the information technology field, addressing the question through a study of reasons for offshoring. It will also examine the strengths and weaknesses of offshoring as well as positive and negative impacts it has had on the information technology field.

History and Background of the Problem

Offshoring is a "strategic use of outside resources to perform activities traditionally handled by internal staff and resources" (Handfield 2006). Offshoring is efficient; it allows an organization to contract major functions to a special facility and reduce production costs (Warren 2002; Nayak et al. 2007). Organizations have long used subcontracting to handle their overflow work or specialty needs jobs. However, the difference between supplementing resources and offshoring is that offshoring involves the restructuring of business activities, specifically the transfer of staff responsibilities to a specialty organization that focuses mainly on a specific need (Davis et al. 2006). For example, a company may offshore a direct mail ordering function but still keep order production and fulfillment in-house. By not having to maintain the staffing as well as computer hardware and software, the company can reduce overhead costs. In addition, the offshore corporation can provide the same service for multiple companies, thus enjoying economies of scale.

Organizations are beginning to strategically shift their information technology needs offshore (Fish and Seydel 2006). From 2002 to 2005, an estimated 160,000+ white collar jobs in the United States were moved to different countries (Kaplan-Leiserson 2004). As labor continues to be inexpensive in Eastern countries, the United States continues to contract work out to overseas businesses. Call center agents and IT specialists are only a few of those losing their job opportunities to India, China, and the Philippines (Kaplan-Leiserson 2004).

Offshoring has been in existence in the United States since the early twentieth century, starting with automakers (Davis et al. 2006). Originally, all parts were constructed domestically; however, as time passed that function was contracted (offshored) to other organizations so the automakers could concentrate on making specialized parts (Davis et al. 2006). Organizations began to experience cost savings from offshoring; however, vertical disintegration soon became the ultimate result of offshoring. Each organization was now producing a subset of the final product (Davis et al. 2006).

Globalization continues to expand the reach of the information technology field, and IT workers in the United States are in constant competition with professionals from other countries who make one-sixth of the U.S. wage (Fish and Seydel 2006). The early stages of offshoring of IT activities during the mid-1900s were focused heavily on hardware in order to cut the high cost of computers and manage them in a controlled environment (Fish and Seydel 2006). It was cost effective to contract out the data processing responsibilities for computers during this time (Fish and Seydel 2006). As technology advanced through the 1990s, organizations began contracting out software development; this would help standardize systems utilizing database management systems.

In the late 1990s, as the U.S labor market became tighter due to the "dot.com" boom and scares about the "year 2000 crisis (Y2K)," a strong shift toward IT and business process offshoring began (Oberst and Jones 2006). As time passed and the IT industry continued to transform, offshoring was the next step for organizations to investigate. Profit-driven organizations looking for cost-saving strategies utilized offshoring as a way to advance their business practices and increase profits. Employees were no longer associated with their jobs; instead, jobs were commodities that could be reassigned to any one or more persons (Oberst and Jones 2006).

Organizations are turning over projects to service companies in the Eastern Hemisphere for extended periods of time. Standard IT functions such as software development and maintenance, network and computer operations, and call centers and help desks are all being offshored (Gibson 2004). As companies continue to find ways to reduce costs and improve their output, offshoring of IT services will remain a sensitive subject.

Organizational Context of the Problem

Offshoring can be defined as an organization's use of products or services from a non-domestic organization (Dutta and Roy 2005). Organizations enter into contracts with the party that will provide their service, with that party being responsible for hiring, training, supervising, and managing personnel (Rubin 1997). As organizations utilize other sources for these services, their contracts specify details of products and/or services included in the contract and what type of service level agreements (SLAs) will be required (Rubin 1997; Chang and King 2005). As organizations look to focus on their core competencies and become more competitive, offshoring becomes part of their corporate strategy (Fish and Seydel 2006).

As the IT field changes, companies work to adapt in order to provide high-quality service, and offshoring becomes a tool for

3

them in adapting to the changes (Lee and Kim 1999). Lacity and Willcocks (1998) stated that organizations have six points of focus when they decide to utilize offshoring services. These points are financial restructuring, core competence, technology catalyst, business transition, business innovation, and new market. These strategic practices are expected to continue as organizations continue to utilize offshoring services.

Originally, as organizations began utilizing their technology, many kept in-house IT departments that focused on system operations, maintenance, and application software development. This has changed as offshoring has become a staple of corporate strategy. Many manufacturers, such as IBM, no longer make hardware or software, as it has become commoditized. Their focus has shifted to support and services to generate more revenue (Davis et al 2006).

The interest in offshoring is a logical step in business as organizations look to improve their revenue and cut their spending. As more organizations start to participate in offshoring their business processes, the subject garners more attention and is strongly being linked to rising unemployment in the United States (Davis 2004). Cost is playing a major role in the decision of these organizations to offshore their business processes. As organizations look for an edge, according to Lacity and Willcocks (1998), they will utilize different cost-cutting forms, such as offshoring of their noncore competent services.

Scope of the Project

This research will focus on offshoring of IT services to foreign countries, rather than domestic outsourcing. Examining trends over the last fifteen years, the researcher will investigate the reasons for organizations offshoring information technology services. This research will focus on the following sections of the information technology field: call centers, help desks, software development, and services.

Definition of Terms

Offshoring, in the context of this research, is the use of foreign organizations to perform information technology services. *Outsourcing*, in contrast, is the use of a domestic organization to perform IT services.

The term *business processes* will be used to describe the sections of an organization—such as a call center, help desk, and IT support—being outsourced/offshored.

Significance of Project

This research will highlight the effects of outsourcing on the information technology field over the past fifteen years. Prior to the early 1990s, technology was not advanced enough for offshoring to be an issue (Fish and Seydel 2006). Offshoring is becoming a common business practice for many IT organizations, affecting several sections of the job market, such as higher education, health insurance, and financial services (Athey and Plotnicki 1998). Universities once had a booming enrollment of students majoring in computer science and management information systems, but they recently have witnessed a steady decline in U.S. students (Athey and Plotnicki 1998). This research therefore has implications for many business sectors and the overall global education and economic landscape.

Literature Review

IT Functions

Organizations are offshoring different functions of IT, mostly lower-level positions that involve repetitive work and minimal critical thinking (Summerfield 2005). The focus is shifting toward offshoring lower-level jobs and putting a higher priority on business knowledge for in-house staff (Dix 2006). Industry studies have shown that only 15 to 20 percent of IT systems are being used by organizations to their true capability, mainly due to a lack of understanding of their IT functions (Madden 2008). Information technology is utilized to create an infrastructure that provides more flexibility and quicker response, and that adapts to an organization's growth and needs (Prouty 2008). Over the past few years, organizations have heavily invested in technology and are richer with data but poorer with information (Sherman 2008). Organizations have IT strategies that are not aligning with their business strategies, causing a separation in strategies (Griffin 2008). Information technology leaders are looking to achieve more with IT, utilizing business integrations and meeting corporate goals (Crockett 2008). Leaders are eliminating the waste of IT by offshoring certain functions in order to contribute to overall business health by improving productivity and product quality and cutting down on production times (Prouty 2008). The interlocking of IT and business strategies can help organizations understand how one section affects the other, and keeps the lines of communication open between both teams (Griffin 2008).

Information technology jobs are becoming harder to fill (Venator 2008); as the lower-level jobs continue to be sent to India and China, IT management positions show no signs of going offshore (PM Network 2007). For the positions such as analysis design, testing and development, and support jobs, there seems to be no end in sight for offshoring (PM Network 2007).

According to Gincel (2005), many of this generation's IT professionals are starting to feel more like the members of the steelworkers' generation. The reliance on offshore partnerships and spread of the global market are causing organizations to change their business models and downsize their IT professionals (Gincel 2005). This, however, is viewed as a positive for IT professionals looking to move up within their IT departments (McGrath 2007). Even though supply might be down, the demand for a new type of IT professional continues to grow (McGrath 2007). Studies showed that, of 140 chief information officers who belonged to the Society for Information Management, two-thirds of them stated that their plan was to maintain or increase the size of their IT staff in 2007 (McGrath 2007). These organizations are looking for responsible decision-making professionals who can involve themselves in strategy development (Crockett 2008).

According to McGrath (2007), one of the fastest-growing occupations of college graduates will be IT jobs; the top five IT jobs are network system and data analyst, software engineer–applications, software engineer–system software, network computer systems administrator, and database administrator. Many of the noncertified skilled jobs and management-level jobs also saw an increase in salaries in 2008 (Eckle 2008). Entry-level IT workers are making way for the demand for IT professionals, such as IT managers (McGrath 2007). According to Gincel (2005), by 2010, 60 percent of people affiliated with IT organizations would be involved in business-facing roles in information, process, and relationships. Information technology workers today who are programmers need to develop new skill sets to manage projects and vendors. Many of these professionals have begun to secure business training in response; the InfoWorld 2005 Compensations Survey indicated that as of 2004, there was a 6 percent increase in the level of MBA applicants from the technology field (Gincel 2005).

Studies show that about 11 percent of the entire IT workforce is made up of IT managers, which represents a 44 percent surge since the dot-com collapse of 2001. About 119,000 IT manager positions have been generated since the 2001 to 2006 time frame, in which over 200,000 programming and technical support jobs have been lost during the same time period (Gincel 2005). Information technology managers, network security managers, wireless network managers, business intelligence analysts, and SAP application developers are also some of the other positions that are not being outsourced (Eckle 2008). In 2007–08 executives and middle managers saw a gain in salary increase, with the mean salary climbing up more than 1 percent; the above-mentioned positions saw an average 20 percent increase in salary (Eckle 2008).

Therefore, U.S. workers who are willing to adapt into managerial roles and offshoring specialists will find themselves with many options in the IT world (Gincel 2005). Many of the new jobs being created in IT involve product architecture and innovation; these managerial positions require strong project management skills (Gincel 2005). In the global economy, the presence of offshoring providers requires that organizations have decision makers to consider the strategic importance of the business strategy (Hickey 2005). Information technology workers are beginning to adapt to the change in the industry; according to Gincel (2005), 19 percent of senior managers have reported having an MBA, a 6 percent increase since 2004. In 2010 the number of senior managers with MBA's has increased to 60 percent (RA Business 2010). In a global IT world, workers need to have more than just technical skills; a "business mind" is required (Venator 2008).

What is Outsourcing?

According to Brooks (2006), outsourcing is described as the utilization of external sources to complete single or multiple tasks in an organization. Hiring another organization to perform the tasks that are normally done in-house is the practice of outsourcing (Bednarzik 2005). Technology has become a large part of how firms achieve success (Benamati and Rajkumar 2002). The ability to transfer all or part of an organization's IT functions has become a strategic plan for many organizations (Benamati and Rajkumar 2002). The continued growth of IT and the difficulty of managing this vital resource are two of the strategic reasons driving outsourcing (Benamati and Rajkumar 2002). An integral process of how IT functions are accomplished requires that an organization rely on outsourcing (Brandel 2004).

An organization has many options when choosing to outsource; it is vital that they select the proper functions and location in order to be successful (Brooks 2006). Organizations today have different options available to them to shift and alter the structure of different departments and functions of their organization (Brooks 2006). For example, an organization can choose to adapt a strategy of allocating over 80 percent of its IT budget to external vendors and be considered a total outsourced organization, or it can choose to outsource selective sections of its IT function to a third-party vendor and be considered a selective outsourcer (Brooks 2006). Companies can choose either strategy and send the functions offshore, producing more of a challenge to how they manage the function (Brooks 2006).

The primary reason organizations outsource is to cut IT costs; however, organizations also look to outsourcing for such strategic reasons as restructuring their IT budget and improving their services (Brooks 2006). Skills and services that are not available internally are often outsourced, allowing the company to focus on core technical functions (Brooks 2006). Reducing the cost of production is one of the benefits gained from outsourcing and leads organizations to choose the strategy (Nayak et al. 2007). Organizations also outsource in order to gain expertise on IT functions that are not part of their core competencies (Brooks 2004). As an organization continues to grow, focusing on core offerings becomes increasingly difficult; therefore, organizations are now looking to outsource IT functions not critical to their competitive edge (Brooks 2004; Kenney 2008). This can be accomplished when a firm can utilize the skills and expertise of their outsourcing firm. Since the firm cannot do everything on its own, it may outsource to gain skill sets that are not available internally (Nayak et al. 2007). A lean and flexible organization is created by outsourcing, providing quick response to its clients and market needs. In such a competitive market, firms cannot take a long time to provide a service or respond to their customers, so keeping the organization lean and flexible makes it quicker to respond (Nayak et al. 2007). According to Brooks (2006), technical issues are cited as another strategic reason for outsourcing. Organizations don't have accessing skills and all technologies internally; outsourcing allows for utilization of those skills and keeps the internal staff focused on core technical functions, improving the delivery of services (Brooks 2006).

Types of Outsourcing

Research shows that various types of outsourcing currently exist (Brooks 2006). Many are categorized into two main types, which encompass all the other categorizations: total outsourcing and selective outsourcing (Brooks 2006). According to Lacity and Willcocks (1998) allocation of 80 to 100 percent of an IT budget to offshore vendors—inclusive of hardware, software, and personnel—is considered to be total outsourcing. Brooks (2006) defines selective outsourcing as assigning hardware, software, or personnel to an offshore vendor, while providing between 20 and 80 percent of the IT budget internally (Lacity and Willcocks 1998).

Studies have shown that selective outsourcing is the favored of the two choices for most organizations. Specifically, 85 percent of organizations are using selective outsourcing and 15 percent are using total outsourcing decisions to obtain the anticipated customer cost savings (Barthelemy and Geyer 2004). The strategy used by organizations of overwhelmingly utilizing selective outsourcing over total outsourcing is based on IT being a changing function (Barthelemy and Geyer 2004). It consists of data centers, application development, and system design and integration, which are candidates for outsourcing to multiple vendors (Barthelemy and Geyer 2004). One of the advantages to selective outsourcing is the ability to spend IT dollars more efficiently to improve IT performance (Weber 2004). Organizations choosing to outsource one or more specific tasks to selected outsourcing vendors can free up their internal IT staff for planning and higher-level work, improve their retention rate in the IT department, maintain control over all their IT intellectual property, and have a greater sense of their maintenance costs (Bowen and

LaMonica 1998). Selective outsourcing encourages collaboration between the IT agency and the vendor in order to improve IT system performance, and it provides improved SLAs on projects (Weber 2004).

Benefits of Offshore Outsourcing

Offshore outsourcing is the "strategic use of outside resources to perform activities traditionally handled by internal staff and resources" (Handfield 2006). As information technology offshore outsourcing continues to grow, more studies continue to be done to help understand why organizations choose to outsource (Xue et al. 2004). These studies have shown that cost savings is one of the benefits that an organization gains from offshore outsourcing (Xue et al. 2004). Cheap labor costs in foreign countries are attractors for organizations sending IT work overseas (Xue et al. 2004). According to Lacity and Willcocks (1998), approximately 80 percent of executives make their offshore outsourcing choice based on IT cost reduction. Surveys in the banking industry have shown that 50 percent of companies with IT expenses of over $5 million were choosing offshore outsourcing as an option to reduce these costs (Ang and Cummings 1997). An estimated 30 to 40 percent cost savings over the course of a project can be gained from offshore outsourcing (Brooks 2004), allowing the organization to gain benefits of cost savings while improving the economic status of countries taking the offshore jobs (Davis et al 2006). Organizations are completing projects that would cost $100 per hour in the United States for $20 an hour in Eastern countries (Overby 2003). The ability to save on producing a product or service allows companies to discount these goods to local consumers, generating growth in volume of goods sold (Davis et al. 2006). Farrell (2005) estimates that in the United States, a savings of $0.58 on the dollar will be gained for every job offshored to India.

While many organizations indicate that cost reduction is the determining factor in choosing to offshore their IT services, other organizations look at offshore outsourcing in order to provide quality of service and competitive advantage (Davis 2004). Evolving consumer needs and demands require organizations to be proactive rather than reactive (Davis 2004). Hiring staff to fulfill customer needs can become expensive for a company, considering the costs of salaries, health benefits, continuing education, and other related expenses, while offshore outsourcing allows for fewer expenses and responsibilities (Ross and Westerman 2003). Research has shown that offshore outsourcing gives organizations a competitive advantage by allowing them to pay less to contractors to perform the same types of jobs originally held in-house (Xue et al. 2004). Organizations now have the ability to produce their products faster and better by offshore outsourcing some of their services and focusing on core competencies (Marks 2003), providing the consumer with a lower-priced product (Mata et al. 1995). Particularly relevant to IT, a reduced time period to provide satisfactory service will help an organization keep a competitive advantage (Nayak et al. 2007).

Organizations are now seeing offshore outsourcing as a business advantage rather than just a cost-saving strategy (Fish and Seydel 2006). The ability of offshore outsourcing companies to now deliver on contracts and become integral parts of an organization's processes is becoming evident as offshore outsourcing grows (Brooks 2006). Organizations continue to turn to offshore outsourcing to keep growing and enhance their competitive edge (Ang and Cummings 1997).

Offshore outsourcing of IT services allows an organization to focus on its core offerings and continue to focus on growing (Brooks 2004). Executives see IT as a non-core activity and often choose this as another reason for offshore outsourcing (Brooks 2006). Companies benefit from focusing on their core competencies; allowing another organization that specializes in IT to deal with all technology issues can be beneficial to an organization (Ross and Westerman 2003; Datz 2003). Information technology offshore outsourcing vendors have a focus around IT management; it is their core business practice, and can they can utilize best practice strategies, unlike their clients (Levina and Ross 2003). The delivery of these services frees up clients to focus entirely on their core competencies (Ross and Westerman 2003). Organizations that benefit from this strategy identify their core competencies before making a decision to offshore their technology needs (Ross and Westerman 2003).

Utilizing an offshore firm to handle IT needs allows a company to benefit from its expertise (Kakumanu and Portanova 2006). Offshore outsourcing as a strategy can improve flexibility of resources and help a firm focus on its business strategy (Fish and Seydel 2006). Offshore outsourcing can provide an organization with a 24/7 technical operation by choosing multiple locations to send IT operations (Kakumanu and Portanova006); offshore outsourcing provides an organization with flexibility by utilizing the rapidly expanding networks of service and backups (Brooks 2004). For instance, organizations that utilize programmers benefit by having views of multiple programmers on any specific project (Kass 2004). Organizations looking for a catalyst in their technology environment can utilize the different ideas and resources provided to them by offshore outsourcing (Lacity and Willcocks 1998).

As organizations continue to look for new markets to showcase their products and services, offshore outsourcing becomes an option (Fish and Seydel 2006). Companies that utilize offshore outsourcing to develop new markets are often the victims of their own success in previous markets (Nayak et al. 2007). Once a market is saturated, firms will be looking for a way to keep growing (Nayak et al. 2007). Offshore outsourcing allows organizations abroad to grow, and that leads to purchase of American products all over the world. As their income increases, the chance is greater that they will be able to purchase these products (Kakumanu and Portanova 2006).

Offshore outsourcing is lowering costs for consumers and manufacturers, creating jobs, and making the economy more efficient (Miller 2004). No longer is it just commodity work that is being sent overseas, with fears of the unknown management challenges or hidden costs for an organization (Brandel 2007). There is less mystery and greater understanding of the pain points associated with utilizing offshoring, allowing organizations to benefit from this strategy (Brandel 2007).

Organizations continue to send more programming and call center jobs offshore, and the topic of offshore outsourcing continues to send waves through the IT world. Workers feel that their jobs continue to move and that no benefit or gain is being provided by outsourcing these jobs (Miller 2004). Studies have shown that the criticism of outsourcing is misplaced, according to Miller (2004); from 1998 to 2003, the spending on offshore outsourcing increased by $7.5 billion, and it was believed that it could reach $31 billion by 2008. In 2009, the offshoring outsourcing contracts signed in the U.S. grew at a rate of 10 percent compared to the 29 percent rate prior to the credit crisis (Lohr, 2009). Global Insight's studies (2004) states that outsourcing is lowering costs and creating jobs. In 2003, 194,000 IT and non-IT jobs were created due to IT outsourcing and by 2008 the number was expected to reach 589,000 (Miller 2004).

Drawbacks to Offshore Outsourcing

According to Bednarz (2004), organizations face several risks when offshore outsourcing their IT functions as compared to keeping them internal to the firm. One of the reasons organizations choose to offshore is to achieve innovation; however, with that innovation comes a lesser degree of quality (Bednarz 2004). In 2005, 21 percent of IT executives surveyed stated that they terminated their offshore outsourcing contracts due to failure to deliver on commitments (Bednarz 2004). Another major issue related to offshore outsourcing is how it impacts domestic employment and the outcome of those effects (Davis et al. 2006). Researchers noted that as of 2005, over 600,000 U.S jobs related to IT had been outsourced (Marks 2003). Drezner (2004) estimates that of the 3.3 million white-collar jobs that will be offshored by 2015, a significant portion will be IT-related. As advances in IT continue, programmers and call center staffers have seen accelerations in layoffs (Mitchell 2004). Research has shown that although there is a fair amount of concern and discontent about the offshore outsourcing wave, corporations are still pushing forward with their plans to offshore their IT services (Gibson 2004).

When transitioning from handling a function internally to offshore outsourcing, organizations often face issues with changes of processes, such as merging of applications with the outsourced staff, or transitioning technical staff (Ross and Westerman 2003). Often the transition is only a small portion of the problem, as organizational change challenges also occur (Earl 1996). Vendors often change the process that is proving problematic or disruptive (Earl 1996). Miscommunication due to lack of understanding usually reduces the benefits of outsourcing a project, because many fail to understand or grasp the new process (Ross and Westerman 2003).

Security becomes an issue as companies begin to send IT functions offshore; organizations should be concerned about entrusting corporate assets to foreign employers (Dineley 2001). Protecting valuable information becomes increasingly difficult when a function is offshored (Kakumanu and Portanova 2006). Evidence exists that companies have lost vital information or have had security breaches of their environment (Kakumanu and Portanova 2006). Potential breaches could have been avoided had technology been kept in-house (Kakumanu and Portanova 2006).

For organizations preparing to take on offshore outsourcing, there are some issues that the less-prepared may face (Adler 2003). Companies that outsource find it very difficult and costly to reverse the decision (Jurison 1995). Barthelemy and Geyer (2004) report that of fifty companies studied, 14 percent deemed their offshore outsourcing experience a failure. The loss of control over their IT resources and projects becomes a worrying factor (Hormonzi et al. 2003). Jurison (1995) also cites another concern for organizations and IT managers: the vendor's inability to deliver. Once an offshore outsourcing contract is signed and in progress, if the vendor cannot deliver on the service level agreements, the contracts are irreversible (Kakumanu and Portanova 2006).

Organizations are also facing a difficult decision in calculating the exact costs of their offshore outsourcing contracts (Purdum 2007). If the overall goal of an organization is to cut costs, offshore outsourcing promises benefits from a cash flow perspective early and in long-term savings (Earl 1996). However, additional care and time needs to be dedicated to the offshore outsourcing of services, as the costs of invoicing, auditing, and communications could rise (Greene 2006). Purdum (2007), states that the problem faced by many purchasing professionals is the inability to look beyond the sticker price, which can handicap an organization and result in under budgeting. Selection of an outsource vendor can cost a company from 0.2 to 2 percent of the actual cost of the deal (Purdum 2007). Essentially, a company doing a $10 million offshore outsourcing deal could spend from $20,000 to $200,000 each year in vendor selection costs (Overby 2003).

Another risk associated with offshore outsourcing is that of a political uprising or war in the country handling services (Davis et al. 2006). Though businesses would like to operate in politically stable countries, cheap labor costs can tempt an organization to choose a less stable country (Davis et al. 2006). Many organizations have most of their applications development and maintenance done out of India, and with pressures in that region companies face potential problems.

Choosing to focus on short-term cost reductions for a project without proper planning or consideration of long-term impacts and changes in business requirements or offshore market conditions tends to bring inadequate results and often leads organizations to be victims of the seven deadly sins: pride, sloth, greed, extravagance, envy, gluttony, and anger (Fox 2003).

Conclusions of the Literature Review

Offshore outsourcing of information technology services is increasing (Benamati and Rajkumar 2002). As organizations continue to take advantage of offshore outsourcing, demand will force prices higher, and companies will find offshore outsourcing less beneficial (Kakumanu and Portanova 2006). Organizations looking to reduce costs and penetrate new marketplaces face different risks by electing to offshore outsource (Earl 1996). If cost reduction is the main goal behind offshoring, companies benefit from early cash flow and long-term cost savings (Earl 1996). However, setup and redeployment costs are overlooked (Earl 1996). Developing a relationship or maintaining an existing relationship with a vendor can still prove to be a lengthy and expensive step (Overby 2003). As has been reported by the IT Outsourcing Institute, roughly 50 percent of organizations that take part in outsourcing do see a reduction in their operating costs (Brooks 2006).

According to Barthelemy and Geyer (2004), many believe that IT outsourcing provides the immediate access to state-of-the-art technology, and lower costs due to vendors pooling demands from different clients. Along with hidden costs, a concern is security threats (Kakumanu and Portanova 2006). Offshore assets are much more difficult to monitor and control; hence they pose a greater security risk to an organization (Kakumanu and Portanova 2006). Information technology outsourcing also brings hazards and risks, from the lack of control and organizations being taken advantage of by opportunistic vendors (Barthelemy and Geyer 2004). According to Madden (2008), as analysts continue to predict a downturn in the economy, organizations need to investigate their outsourcing contracts and make sure they are valuable from a cost savings perspective.

Organizations continue to focus their attention on outsourcing in different forms: total and selective (Brooks 2006). Total outsourcing forces the organization to send out its entire IT budget, while selective outsourcing requires it to spend only some of its IT dollars (Brooks 2006). As organizations continue to rely on offshore partnerships and the global marketplace continues to spread, companies are being forced to reshuffle and downsize their IT departments (Gincel 2005).

According to Davis et al. (2006), offshore outsourcing has an impact on domestic levels of employment. Cheap labor costs in India are winning over organizations that are short on resources in the United States (Bednarz 2004). Drezner (2004) states that 220,000 jobs a year will be lost due to offshoring over a fifteen-year period, a number that could continue to increase as organizations acknowledge the negative effects of offshore outsourcing, but continue to utilize the services (Gibson 2004). The shift to offshore IT continues to grow at a rapid pace, and as organizations continue to look for new ways to reduce costs and gain competitive advantages (Bednarz 2004), IT workers will continue to see acceleration in layoffs (Mitchell 2004).

Methodology

Summary of Project Purpose

Information technology offshoring has become a standard in doing business in the United States, and the purpose of this research will be to investigate if and how offshoring affects the IT field. Information technology offshoring, according to studies, has negative impacts on the industry, due to loss of jobs, lack of security, reduced quality of products, and loss of control and miscommunication due to lack of understanding (Bednarzik 2005).

Organizations, however, continue to send jobs offshore, choosing to totally outsource or selectively outsource. The options of both give organizations the ability to control what goes out and what comes in. Outsourcing gives organizations a cost saving strategy, utilization of outside expertise, improved focus on their core competencies, and development in new markets.

This research investigates the effects of offshoring on the information technology field, looking at the perspective of benefits and drawbacks of offshoring. Outsourcing provides organizations with many advantages in order to move them forward in utilizing technology; the drawbacks are evident, however, and this research will investigate both views.

Data Collection and Presentation

Data were obtained from a Robert Half Technology industry survey, and studies included were done by Matt McGarth and the U.S. Department of Labor, from 1994 to 2008. These surveys and studies were chosen because they showed consistent data collection over the past ten years on the information technology field and provided numerical qualification of the effects of offshoring on the market. Robert Half Technology provides information technology consultants in all parts of the United States, and it publishes the *Salary Guide* yearly on positions in the information technology field.

Several charts were created or identified from the data from the above sources to show the effects of outsourcing on the IT field. A chart format was chosen because of the need to organize diverse groups of data in a concise form. Utilizing charts provides immediate indication of relationships between data items and allows comparison across multiple items within the same group.

Chart 1 compares the top six IT jobs in the United States, analyzing their growth since 2004 and their compensation growth since 2007. Chart 2 shows the percentage distribution of IT-sector jobs. This chart compares the high-wage positions to the low-wage positions in the IT field from 2000 to 2004. Chart 3 compares the manufacturing and services roles in information technology from 1994 to 2004. Information technology sector employment from the year 2000 to 2004 is compared in Chart 4, showing the types of jobs that were filled during those years. Chart 5 depicts the locations where the IT jobs are being sent, and Chart 6 shows the compensation that these employees are getting at these jobs. Chart 7 shows the numbers of layoffs in IT departments across the United States from 2001 to 2008. This covers the job losses on the hardware and software side. Chart 8 shows the findings on how outsourcing has affected the U.S. job market since 2000. The X-axis indicates the years, and the Y-axis indicates the numbers of jobs lost in the thousands. Chart 9 shows the employment in the IT sector broken down into different positions, from 2000 to 2004.

All the charts help show the IT field from different perspectives, showing compensation, layoffs, top jobs, and skill sets in the field, and even offshoring locations, giving an insight into the effects of outsourcing on the IT field in America.

Research Method and Design

As the data for the research come from diverse sources and encompass many areas of study, a combination of qualitative and quantitative methodologies was needed. Historical-comparative research methodology was employed in this research. Historical-comparative research is a preferable method for assembling and analyzing large amounts of diverse secondary data, and it can be used to examine both qualitative and quantitative studies addressing a single research question or topic (Neumann 2008).

In historical-comparative research, tables, charts, and other graphic representations of data are created and/or taken from source information. These graphics are then analyzed and used in series to build an argument by addressing the various aspects or facets of the research question explored (Neumann 2008).

Historical-comparative research is particularly useful in research in which the research question is too complex to be addressed in a single empirical study, but instead requires consideration of multiple studies (Neumann 2008).

The nine charts mentioned above were used in the analysis of the research problem. Three of the charts—1, 5, and 6—were created by the researcher using data from the sources indicated. The remaining charts—2, 3, 4, 7, 8, and 9—were taken directly from source materials as indicated. Of note, Chart 7 is not presented in the findings in its entirety; only relevant data are included. The complete chart as published appears in Appendix 1.

The charts used in this book, were arranged by topic rather than chronologically, as the approach was more revealing in the analysis.

Findings

Data were obtained from a Robert Half Technology industry survey, and studies included were done by Matt McGarth and the U.S. Department of Labor, from 1994 to 2008. Several charts were identified in the findings to show the effects of outsourcing on the IT field. A comparison was done amongst the top six IT jobs in the United States, analyzing their growth since 2004 and their compensation growth since 2007. Another chart was reviewed showing the percentage distribution of IT-sector jobs. This chart compared the high-wage positions to the low-wage positions in the IT field from 2000 to 2004. The distribution of the employment was also compared in this section to show the trend in manufacturing and service jobs in the IT field over the ten-year period of 1994 to 2004.

Other findings compare the locations where the offshored jobs were heading and the compensation employees in those parts of the world were receiving. That chart will give insight into the number of jobs being distributed to India, and specifically the different cities in the country. A U.S. Department of Labor chart was studied to show the number of layoffs in the information technology sector from 2001 to 2008, specifically comparing the hardware layoffs to the software and services layoffs. Though the data from the third and fourth quarter layoff could not be obtained due to the time frame of this thesis, this chart helps show several misunderstandings about the IT layoffs related to offshoring. One of the most significant charts in the findings compares the number of U.S. job lost due to offshoring since 2000. This chart is a graph comparing the jobs lost in the United

States and the number of jobs, IT- and non-IT-related, that have been created in that time frame.

All the charts help show the IT field from different perspectives, showing compensation, layoffs, top jobs, and skill sets in the field and even outsourcing locations, giving an insight into the effects of outsourcing on the IT field in America.

Chart 1. Top IT Jobs

The chart below shows the top six in-demand jobs in information technology and their growth from 2004 to 2008. These jobs are related to the software or services section of information technology.

Job Title	Growth Since 2004	Skill Set	2007 Compensation	2008 Compensation	Growth From 2007 to 2008	Location
IT Manager/Project Manager	44%	Project Management	$72,500 - $106,250	$76,500 - $111,500	5.0%	New England
Network Sytems & Data Communication Analyst	54.60%	Network Administration (Cisco)	$61,250 - $86,500	$64,250 - $91,750	5.6%	Mid-Atlantic
Software Engineer - Applications	48%	ERP Implementation	$55,250 - $90,750	$57,500 -$96,750	6.0%	South-Atlantic
Software Engineer - Systems Software	43%	.Net Development	$71,000 - $102,000	$76,250 - $108,250	6.6%	Pacific
Network Computer Systems Administrator	38.40%	Wireless Network Management	$65,750 - $90,250	$67,250 - $93,500	3.0%	East, South Central
Database Administrator	38.20%	Database Management (DB2, Oracle, SQL)	$84,750 - $116,000	$88,750-$122,750	5.4%	West, North, Central USA

(Robert Half Technology Survey 2007)

From 2004 to 2008, these jobs averaged a growth of 30 percent as more organizations look for IT professionals who will provide business value to their organizations. Significant growth has come from the data communication analyst position, showing a 54.6 percent growth since 2004 and a compensation increase of 5.6 percent. A majority of data communication analyst jobs are found in the Mid-

Atlantic United States. These job positions saw a very steady increase in compensation from 2007 to 2008. Software engineers for system software saw the most increase in salary, showing a 6.6 percent increase. As organizations continue to use business intelligent software to provide more value for their businesses, this position continues to grow. These jobs are spread all across the country, and they are providing high compensation due to the business value that they are bringing. The jobs listed above require business-oriented skill sets, and bring a managerial aspect to the position, hence making them tougher candidates for outsourcing. According to McGrath (2007), the IT worker of today is one who needs to possess technical skills and the business skills to manage and analyze projects.

Chart 2. Wage Distribution

Chart 2 depicts the wage disparity between the levels of IT jobs. According to the 2004 Monthly Labor Review done by the U.S. Department of Labor, high-wage jobs are classified as information systems managers, software programmers, computer systems analysts, software engineers, and network computer system administrators and analysts. Low-wage jobs are classified as support specialists/ help desk specialists, computer operators, data entry operators, and auto-teller and office machinery repairers.

Percent distribution of IT-Sector by high and low wage occupations 2000 -2004		
Year	High-wage(1)	Low-wage(2)
2000	64.4%	35.6%
2001	66.5%	33.5%
2002	66.9%	33.1%
2003	68.9%	31.1%
2004	69.1%	30.9%

(Bednarzik 2005)

The chart shows that between 2000 and 2004, there were increases in the high-wage jobs and steady decreases in the low-wage jobs. Many of the IT professional jobs are having a bigger business impact in their organizations; hence the reason for the increase in wages. Information technology professionals have been forced to enhance their skill sets to include more business-related capabilities (McGrath 2007). As organizations continue the search for business value positions, the IT field continued to see a growth. Information technology managers, database administrators, and application developers are helping bring business and IT together.

Low-wage jobs continue to decline as they become the focus of outsourcing; these jobs do not require any business capabilities, and organizations are finding them to be candidates for selective outsourcing. Support specialists/help desk specialists, computer operators, and data entry operators are jobs that can easily be outsourced due to the simplicity of the positions. These positions don't show any business value to the organization, nor can they bridge the gap between senior level management and IT, so they will continue to be candidates for offshore outsourcing. As these positions get outsourced, the organizations that do decide to keep them will continue to shrink the wages and eventually push them offshore or do away with the position.

Chart 3. IT Sector Employment in Manufacturing and Services

This chart depicts the reductions in employment in IT manufacturing and the increases in the services aspect of IT from 1994 to 2004. The chart compares both aspects of IT and shows the changes over that time period.

Percent distribution of IT-Sector employment in manufacturing and services, 1994-2004		
Year	Manufacturing	Services
1994	66.6%	33.4%
1995	64.5%	35.5%
1996	62.6%	37.4%
1997	60.1%	39.9%
1998	57.1%	42.9%
1999	52.7%	47.3%
2000	50.5%	49.5%
2001	49.0%	51.0%
2002	47.8%	52.2%
2003	46.4%	53.6%
2004	45.4%	54.6%

(Bednarzik 2005)

This chart shows that from 1994 to 2004 the IT job growth from the manufacturing side was on a steady decline, and on the services side an increase was noticed in the early 1990s until 2001, when the economy was in a recession. It has been difficult to draw a conclusion as to the effects on offshoring during the recession period due to the business cycle change (Bednarzik 2005); however, this chart helps convey some conclusions. In 1994 the IT services jobs were at 33.4 percent and by 2000 they were up to 50 percent. In the years 2000 to 2004, an increase of 4.6 percent was noticed in the IT services sector of the job market, showing that not as much outsourcing was occurring. Organizations were beginning to see the value of the service-based position and began to hire more IT professionals to fit the positions of IT managers, data communications analysts, and network administrators. Manufacturing jobs in the IT sector continued to decline, from 66 percent in 1994 to 45 percent by 2004. Chip designers Intel and AMD continued to push their jobs overseas as the cost of labor in outsourcing countries continued to be cheaper. As the focus of organizations changed to looking for IT professionals, services jobs increased as manufacturing jobs decreased, as is indicated on the chart.

Chart 4. IT Sector Employment

Chart 4 compares the employment in the IT sector by occupation from 2000 to 2004, showing trends in the types of employment during these years in the thousands.

		Employment in the IT sector, by occupation, 2000-2004				
		Occupation	2000	2001	2002	2003	2004
Total - IT Sector			4718	4795	4510	4494	4495
Computer and Info System Managers			228	316	323	347	337
Computer programmers			745	689	630	563	564
Computer & Information scientist and system analysts			835	734	682	722	700
Computer Hardware engineers			83	100	76	99	96
Computer Software engineers			739	745	715	758	813
Computer Support Specialists			350	355	353	330	325
Database administrator			54	66	84	72	94
Network and Computer systems administrators			154	185	179	176	190
Network systems and data communication analysts			305	353	328	259	312
Computer Operators			313	324	283	191	191
Data entry keyers			632	623	542	581	504
Computer auto-teller and office machine repairers			280	305	315	296	369

(Bednarzik 2005)

This chart shows the types of jobs in the IT field that were filled from 2000 to 2004. Information technology sector jobs showed an increase of 49 percent from 2000 to 2001. A decline occurred in 2002; however, growth was reestablished from 2003 to 2004. Low-level jobs such as programmers, support specialists, and data entry operators all declined from 2000 to 2004, with programmers seeing the most hit of the three, at a 24 percent decline from 2000 to 2004. This chart further goes on to show that as the low-level jobs decreased, the high-wage jobs grew. From 2000 to 2004, information system managers saw a growth of 47.8 percent, database administrators saw a jump of 74 percent, and network systems administrators saw an increase of 23 percent. Organizations continued to see value in these positions, so they hired more to fulfill business needs rather than technical needs.

Chart 5. Outsource Locations

The fifth chart compares the locations in India where different focus jobs are being placed. The chart shows the focus jobs, locations, and the numbers of workers who have been hired since 2003.

Focus	Outsource Location	Number of Workers in 2003 (in Thousands)
Back Office	Mumbai (Bombay)	62,050
Call Center	Bangalore	110,000
	Kolkata (Calcutta)	7,300
Chip Designer	Bangalore	109,500
Programming	Pune	7,300
Software	Delhi	73,000
Total		369,150

(Fox 2003)

The focus of jobs being outsourced as of 2003 shows a majority were in the call center section of IT, closely followed by the manufacturing sector of IT. Dell specifically led the charge in sending call center jobs to Bangalore and Kolkata, India. In 2003, the cities have a combined total of 117,300 low-level jobs filled. Chip designers alone, such as Intel, sent over 109,500 jobs, focusing mainly on the Bangalore area of India. The infrastructures have been built from a technology standpoint in these cities, and they are allowing companies like Intel, Cisco, Oracle, and Phillips to outsource their low-level jobs and gain cost savings. In total, 369,150 jobs were outsourced to the sections of India mentioned above; Bangalore led the way with 59.4 percent of those jobs. Bangalore is considered the Silicon Valley of India, being 3,000 feet above sea level and boasting some of the best education in the country; it is the ideal location for organizations to send their outsourced positions.

Chart 6. Compensation Based on Location

This chart focuses on the compensation growth received by employees in the different positions in the outsourced destination. It also takes a look at the numbers of positions that are available in the country.

Focus	Outsource Location	Number of Workers	Position	Compensation Growth since 2005
Programming	Pune	7,300	Entry Level	10% - 12%
Chip Designer	Bangalore	109,500	Entry Level	10% - 12%
Call Center	Kolkata (Calcutta)	7,300	Entry Level	10% - 12%
Back Office	Mumbai (Bombay)	62,050	Entry Level	10% - 12%
Software	Delhi	73,000	Middle Management	15% - 20%

(Fox 2003)

As is evidenced by the chart, the jobs being outsourced to India are entry-level positions. Since 2005, 186,150 entry-level jobs have been sent over to India, with only 28 percent of the total jobs being middle management positions. This chart demonstrates that many of the middle management positions are staying onshore and organizations are pushing out their low-level, entry-level positions.

The compensation growth on these jobs has remained constant between 10 and 12 percent for the entry-level position and 15 and 20 percent for the middle management position. An organization pursuing the outsourcing strategy will still pay less to hire someone in Bangalore, India, to design a new processor chip, than to hire someone in the United States.

Chart 7. IT: Extended Mass Layoff Events and Separations

Chart 7, provided by the U.S. Department of Labor, studies information technology layoffs from 2001 to 2008, and shows conclusively that IT jobs are not in jeopardy, as many believe. This amended chart compares computer hardware layoffs to software and services layoffs.

Year	Total Extend Mass Layoffs		Computer Hardware (2)		Software and computer Services (3)	
	Layoff Events	Separations	Layoff Events	Separations	Layoff Events	Separations
2001						
Total	7375	1524732	503	92587	196	28512
2002						
Total	6337	1272331	303	59653	135	18164
2003						
Total	6181	1216886	196	32689	78	13194
2004						
Total	5010	993909	76	11524	52	8575
2005						
Total	4881	884661	75	11928	32	5667
2006						
Total	4885	935969	48	12036	23	3503
2007						
Total	5363	995935	72	13379	19	2791
2008						
First quarter	1340	229870	19	3040	9	987
Second quarter	1534	299886	19	3181	5	618

(Bureau of Labor 2008)

From 2001 to 2006 a steady decline in overall IT layoffs was seen, starting at 7,375 and dropping to 4,885 by the year 2006, a 66 percent decline. After a down period in 2006 to 2007 when layoffs rose by 478 jobs, the first and second quarters of 2008 showed the highest first and second quarter layoffs in IT since 2003. Further analysis shows that fewer layoffs were observed in the software and services sector of IT. This chart shows that of the 34,669 jobs laid off from 2001 to 2006; only 516 were software and services related; in comparison, 1,201 were hardware related.

In 2001, 2.6 percent of the layoffs were related to the software and services fields, and by 2007 that number, according to the charts, had dropped below 1 percent. Information technology professionals, such as IT managers, database administrators, and network administrators, were not seeing as many layoffs as those involved in the construction of the technology parts. In 2001, 7.9 percent of the total job layoffs were related to computer hardware; by the year 2006 that number showed a significant decline but not at the same rate as computer services.

As the first and second quarters of 2008 drew to a close, the same trend continued to show up. First and second quarter hardware layoffs showed a 1.4 percent and 1.2 percent layoff rate, respectively, while software and services layoffs showed a 0.67 percent and 0.32 percent layoff rate, respectively.

This chart shows that outsourcing of low-level IT jobs continued to grow and the more business-focused positions were being kept in-house. As organizations looked to give themselves a competitive advantage, they realized that IT professionals with technical and business skill sets were needed in their companies to take them to the next level, and they were filling those positions steadily.

Chart 8. U.S. Job Loss Due to Outsourcing

This chart shows the findings on how outsourcing affected the U.S. job markets during 2000 to 2008. The X-axis indicates the years, and the Y-axis indicates the numbers of jobs lost in the thousands.

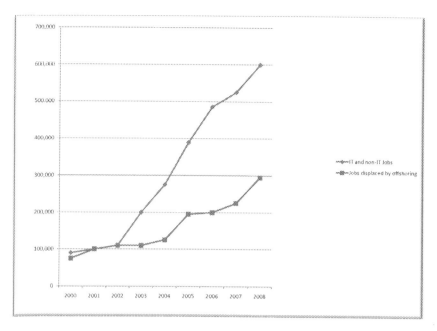

(Miller 2004)

From 2000 to 2008, IT software/services job loss climbed from 50,000 to just under 300,000 jobs. The job displacements leveled off at roughly 300,000 in 2008 as per the chart. As the trends continued, the jobs have leveled off since 2008. However, the red line indicates that new IT and non-IT jobs created by outsourcing have grown from 50,000 in 2000 to roughly 600,000 thousand jobs in 2008. This chart also indicates that if outsourcing continues to increase, the number of jobs created will increase. According to this chart, the number of jobs created will be double the number of jobs lost in 2008. Rather than reducing the number of jobs in the United States, offshore outsourcing is lowering costs for everyone and creating more jobs, because the economy has become more efficient (Miller 2004).

This section has compared the quantitative data from different sources to show the effects of offshoring on the IT field. Comparing the growth of the top six jobs since 2004 and the compensation growth that these jobs are receiving has allowed an insight as to the hiring trends of organizations. To further show the hiring trends, charts have been provided to show the layoffs and compensation of high-level and low-level jobs from 2000 to 2008. These charts have shown that outsourcing is not affecting the IT field as significantly as expected.

Chart 9. IT Sector Employment

This chart shows the thousands of employment in the IT sector broken down into different positions from 2000 to 2004. This chart will help analyze the breakdown of low-level and high-level jobs.

Employment in the IT sector, by occupation, 2000-2004							
Occupation			2000	2001	2002	2003	2004
Total - IT Sector			4718	4795	4510	4494	4495
Computer and Info System Managers			228	316	323	347	337
Computer programmers			745	689	630	563	564
Computer & Information scientist and system analysts			835	734	682	722	700
Computer Hardware engineers			83	100	76	99	96
Computer Software engineers			739	745	715	758	813
Computer Support Specialists			350	355	353	330	325
Database administrator			54	66	84	72	94
Network and Computer systems administrators			154	185	179	176	190
Network systems and data communication analysts			305	353	328	259	312
Computer Operators			313	324	283	191	191
Data entry keyers			632	623	542	581	504
Computer auto-teller and office machine repairers			280	305	315	296	369

		2403	2396	2199	2060	2049
% of Low Level		51%	50%	49%	46%	46%
		2315	2399	2311	2334	2446
% of High Level		49%	50%	51%	52%	54%

(Bednarzik 2005)

Information technology sector jobs overall have seen a steady decline since 2001 according to this chart, with most of that coming from the numbers of jobs being outsourced. However, the numbers of jobs that have been outsourced since 2000 have depended on the occupation within the IT sector. Analysis of this chart shows that low-level jobs were at 51 percent in 2000, and slowly dropped to 46 percent in 2004, a 5 percent decrease in the number of data entry, computer operator, and hardware engineer jobs. High-level jobs, however, showed a 1 percent increase from 2000 to 2001 and then increased another 4 percent from 2001 to 2004. The chart shows that the IT sector was making the change from low-level positions to business-focused roles. The network system analyst position showed the most growth, 98 percent, from 2000 to 2004, followed closely by the software engineer and then the network administrator. All these roles are high-level positions that require technical knowledge and understanding of the business in order to bring value. This chart shows that the growth from 2000 carried on through the events of September 11, 2001, and continued to grow until 2004.

Conclusions and Recommendations

Interpretations of Findings

Employment trends categorized by industry and occupation suggest that offshoring in the information technology sector occurs, but historically has not had significant impacts beyond those on entry-level positions (Bednarzik 2005). The nature of the problem was the focus of this research, and the findings section leads to the conclusion that in fact the impact of offshoring or outsourcing on the information technology sector affects mostly low-wage jobs such as chip design, programming, software development, and call centers. Many of the roles in information technology that are business focused are remaining onshore.

Organizations realize the importance to their IT department of having positions that help bring value to their business, and the roles that are being left onshore are continuing to grow year to year. These roles, such as IT managers and database administrators, have seen an average growth of 30 percent since 2004, and from 2007 to 2008 there was an average 5.3 percent increase in compensation for these roles.

Experts believe that the reason most organizations outsource is to achieve innovation and find better ways of solving their IT needs; however, as business professionals continue to emerge with advanced degrees, they bring their own innovation into these roles and change the face of organizations. Organizations are looking to the business professionals to help provide them with competitive advantages over their rivals, as they look to increase their market share. According to Griffin (2008), many organizations have IT strategies that are not aligning with their business strategies, causing a separation in strategies. However, the findings from this research indicate that there is a rapid growth in the IT professional positions as organizations continue to put their business needs in the forefront of their IT strategies. Industry studies showed that 15 to 20 percent of their IT systems are being used by organizations to their true capability (Madden 2008), and the increase in the employment of IT professionals since 2000 has grown nearly 20 percent. Organizations recognize the need for their IT systems and are willing to compensate their employees for their knowledge and business value.

Furthermore, the wage increases correlate with the demand for these professionals; organizations understand that for the business value that these professionals bring, compensation must be matched. From 2000 to 2004, high-wage IT jobs increased 5 percent, as the professionals continued to have greater impacts on their business. The low-wage jobs, such as support/help desk specialist, computer operators, and data entry operators decreased by 5 percent during this time period.

The findings of this research indicate that from 2000 to 2004, as the manufacturing sector of IT continued to decrease the number of low-wage employees, the wages increased for high-wage jobs in the sector. Those low-wage jobs are the jobs that are being outsourced; hence the significant change in wages.

Several researchers have stated that as of 2005, over 600,000 U.S jobs related to IT had been outsourced (Marks 2003); from 2005 to 2007, however, the number of IT layoffs related to management and services roles, was just under 1 percent according to the U.S. Department of Labor Monthly Review. This research also found that from 1994 to 2004, the number of jobs lost in the United States related to IT from a manufacturing perspective was 21 percent; however, from a management and services perspective, there was a 21 percent increase. There is ample evidence that outsourcing is affecting the U.S IT job market. However, in the IT sector it is for low-level jobs, while the more business-facing roles are staying onshore and growing significantly.

As was concluded in the literature review, many experts believe that the security and the poor quality of work are among the drawbacks of outsourcing. Findings, in contrast, indicate that since 2003 the number of low-level jobs going offshore is increasing. In 2003 alone, over 109,500 chip designer jobs were sent to India, focusing on the well-educated cities such as Bangalore. Outsourcing to these specific locations is very inexpensive, so the findings show that since 2003 an increased percentage of low-level IT jobs continue to be sent over. The total number of jobs sent over to specific locations like Bangalore, Kolkata, and Mumbai totaled 369,150 jobs. The lifestyles in these places also are being significantly changed. Since 2005, a growth of 10 to 12 percent has been seen in compensation for workers in these locations, showing that an increase in compensation could change the cost saving mentality that organizations have with outsourcing. As the lifestyles continue to change in Delhi, Kolkata, Mumbai, and Bangalore, the compensation will increase and that cost will be passed on to the organizations. Experts have stated that approximately 80 percent of executives state that their outsourcing choices are based on IT cost reduction (Lacity and Willcocks 1998). It is important to note that, if compensation in those parts of India continues to grow, those cost reductions could diminish.

Drezner (2004) estimated that 3.3 million white-collar jobs will be offshored by 2015, and a significant portion will be information technology related. However, findings from this research show that from 2000 to 2008, approximately 300,000 jobs were lost, showing an average of 37,500 each year. In contrast, from 2000 to 2008, approximately 600,000 IT and non-IT jobs were created due to outsourcing, showing an average of 75,000 a year. If this rate continues, by 2015 an additional 525,000 jobs, IT and non-IT, will be created. This means that the fear of the number of jobs being lost is arguable, as the findings show that there has been a surplus in the numbers of IT and non-IT jobs during that time. Organizations realize that outsourcing is part of doing business; however, they have kept the important IT positions onshore in order to have a competitive advantage.

Many organizations understand the importance and benefits of outsourcing, so they are taking advantage of the inexpensive labor and shipping the jobs out overseas; however, they have adopted a selective outsourcing method. Selective outsourcing allows organizations to send jobs such as call centers, chip design, and other low-level jobs to different parts of India, or other nations, and keep the IT professional positions and business-facing roles, such as IT managers, database administrators, and project managers, in-house.

Conclusions

The findings of this study clearly show that outsourcing does exist, but the manner in which it does, is not as damaging as many have feared. If low-level jobs such as chip design, call centers, and software design are the positions being outsourced, then all is not lost for the American IT professional. The charts in these findings show that as the low-level jobs are being outsourced, the high-level jobs are increasing. Specifically, Chart 2, the distribution of IT sector chart, shows the decrease in manufacturing jobs and the increase in services jobs. This chart shows that companies are clearly using selective outsourcing methods, as they are keeping their highly skilled professionals in-house and continue to grow that market. Organizations are starting to see chip design, data entry, and call center support as IT positions of the past and have started to embrace the new IT professional, with the technical expertise and business savvy to help them achieve their business goals and focus on their core values. These services and management positions have grown year by year, and comparison to Chart 1, the top IT jobs chart of 2007 shows that the growth continued in that sector of the IT field. The rate at which the jobs grew since 2004 is much more significant than from 2000 to 2004. In 2010, Network administrators, IT managers and Software engineers are classified as the top jobs in the IT field (Nuwave 2010). This finding leads this researcher to believe that organizations saw the value of these positions and are consistently hiring more IT professionals who will bring in a business mind set to their IT departments and organizations.

In addition, compensation has grown drastically for the IT professionals from 2008 to 2010 and that growth will likely continue as organizations see the value that these roles bring. In 2010, the salary increase for IT workers is expected to grow by 1.8 percent (Computer Economics 2010). Though a minimal number over previous years, the increase shows that IT executives are willing to pay to retain their best employees and grow their organization (Computer Economics 2010). Chart 1 complements Chart 2, as it shows that the organizations value the highly skilled business professionals and are willing to compensate them for their knowledge. The 5.1 percent wage increase from 2000 to 2004 and the average of 5 percent increase from 2007 to 2008 show that as organizations continue to understand the importance of these roles, compensation will continue to increase. These high-level jobs exist in all parts of the country, and regardless of the location, compensation and demand for qualified professionals to fill these positions continue to grow.

Chart 7 explains that the layoffs researchers and studies have stated exist in the IT field are related to computer hardware and not the software and services section. This chart shows that from 2001 to 2007, the number of jobs in the services sector of IT decreased, starting at 2.6 percent in 2001 and ending at less than 1 percent by the end of 2007. This chart makes it evident to this researcher that the fears of outsourcing affecting the IT field are exaggerated, as the findings show that while there is outsourcing, it is not affecting the rapidly growing high-wage IT professional positions that organizations need to hold a competitive advantage.

Chart 8 clearly indicates that outsourcing exists and has been growing since 2000; however, it also clearly shows that for every one job outsourced, two in the IT or non-IT field are created. In 2008 roughly 300,000 IT jobs were displaced by offshoring; that same year double the number of jobs were created by the economic activity created by outsourcing. It is clear that the more low-wage jobs are outsourced, the more high-wage jobs are created. The more low-level jobs are shipped to locations like Mumbai and Bangalore, the greater need organizations will have for project managers, IT managers, and database administrators. The fear of many that outsourcing is crippling the IT sector or the economy is not shown to be a viable argument by this research, as Chart 8 indicates. In fact, the jobs lost to outsourcing and jobs created are at a 1 to 2 ratio, showing that even during the downtime in the economy after September 11, more jobs were still created from outsourcing. The chart has grown over the years, with acceleration beginning in 2003 when approximately 50,000 jobs were created over the number of jobs lost.

In conclusion, organizations are looking for ways to gain a competitive advantage, and they have settled on hiring IT professionals who bring not only technical strength to their organizations, but also a business mentality. Many IT professionals are now returning to school to earn master of business administration or other postgraduate degrees in order to be competitive in the job field and provide these organizations with the business minds they are seeking. As is echoed by Chart 1, it is clear that growth exists in these jobs that are in demand by the market. An average of 44 percent growth amongst those four jobs leads this researcher to conclude that outsourcing is not having an adverse affect on the information technology field, as some might be led to believe; on the contrary, it is actually having a positive effect. Outsourcing is doubling the number of IT and non-IT jobs in the market and forcing IT professionals to further educate themselves in order to compete for these high-wage positions. It can be said that outsourcing is actually cleansing the IT sector, by making sure the people with the appropriate skill sets are in the business-facing roles.

Recommendations for Change

This research has shown that organizations can still partake in offshoring or outsourcing, and continue to help themselves and keep a competitive advantage. Selective outsourcing is one of the most noticeable business practices found in this research. It allows organizations to outsource some functions of IT, while still growing the core part of their business. Organizations can utilize this practice in order to help strengthen their advantage against the competition and also to help the IT sector.

Utilizing selective outsourcing will allow organizations to send out jobs such as call centers and low-level programming positions, saving them money. Then they can apply those savings to bringing in IT professionals such as software designer–applications, in order to help design an application that will help run their day-to-day operations better and more consistently, and an IT manager to oversee this position and other positions in their IT function. This strategy also gives an organization more ownership over its outsourcing, and the fears of risk and security breaches can be eliminated.

Organizations should also consider utilizing outsourcing firms in the same cities and locations in order to have a better understanding of their costs and the savings they will realize. One of the drawbacks of outsourcing was the inability to calculate the exact savings that outsourcing provides. However, if an organization utilizes the same company in the same location, such as Mumbai or Bangalore, it can have a better understanding of what it is saving and how it can be applied to other areas of the business.

Recommendations for Further Research

Further research should be conducted on the locations where the IT jobs are being outsourced to. Research in this area could lead to finding more inexpensive locations for organizations to outsource their low-level jobs. This research would also show the compensation growth rate for these countries or cities.

Further research also should focus on how outsourcing will change the state of the economy in 2011 since the recession of 2009. As the economy continues to fluctuate and cutbacks continue to happen, organizations will begin to review their IT needs and determine if outsourcing provides the cost savings and benefits discussed in the literature review section. Studies conducted in 2008 by the TPI index reveal that over 200 outsourcing contracts were signed in 2008 alone, totaling over $80 billion, which was the strongest since 1998. In 2010 the TPI index that 420 outsourcing contracts valued at $15 billion dollars were set to expire, an increase of 40 percent compared to 2009 and is at the highest level in five years (Duvall 2010). These contracts will be broken into pieces and outsourced to different firms, increasing the outsourcing number by 9 percent since 2009 (Duvall 2010). Organizations are looking for a way to cut expenditures and become more profitable. A study done in January of 2007 shows that from 2007 to 2010, there will be a $19 million increase in IT services, a $5 million growth from 2003. Recent studies conducted by Market Research Media, puts IT spending at $533 billion dollars from 2010 to 2015. The increase in services spend by the end of This growth rate is seen only in services and not in the manufacturing segment, where cutbacks are continuing to increase.

Extensive research could show the exact types of jobs that are still being outsourced. Manufacturing and call center jobs have been the main outsourcing candidate for the past ten years. As of 2010, offshoring is still considered the reason for the slow economic growth and the slowdown of IT job recovery (Thibodeau 2010). The economic downturn and poor performance of the global market has not reduced the offshoring efforts of organizations. According to Thibodeau (2010), the US.S Department of labor reported that IT employment increased in the November 2010 by 600 jobs, on the base of 3.9 million jobs. Of the 39,000 jobs added in November of 2010, only 600 of those jobs were IT related. Studies conducted by research groups of organizations with over one billion dollars in revenue, concluded that by 2014, lower-level IT jobs will be down by fifty percent from four million jobs to two million jobs (Thibodeau 2010).

Bibliography

Adler, T. R. (2003–2004, winter). Member trust in teams: a synthesized analysis of contract negotiation in outsourcing IT work. *Journal of Computer Information Systems,* winter 2003–2004, 6–16. Retrieved from Business Source Premier database.

Allen, T. D., Freeman, D. M., Russell, J. E., Reizenstein, R. C., and Rentz, J. O. (2001). Survivor reactions to organizational downsizing: does time ease the pain? *Journal of Occupational and Organizational Psychology*, 145–164. Retrieved from ProQuest database.

Ang, S., and Cummings, L. L. (1997, May/June). Strategic response to institutional influences on information systems outsourcing. *Organization Science*, 8(3). Retrieved from Business Source Premier database.

Annesley, C. (2006, September). Offshoring offers greater IT flexibility. *Computer Weekly*, 56.

Athey, S. and Plotnicki, J. The evaluation of job opportunities for IT professionals. *Journal of Computer Information Systems* 38(4), 1998, 71-88

Avery, S. (2004, August 24). Outsourcing harness strengths of suppliers. *Purchasing,* 129(3). Retrieved from Business Source Premier database.

Barthelemy, J., and Geyer, D. (2004, spring). The determinants of total IT outsourcing: an empirical investigation of French and German firms. *Journal of Computer Information Systems*, 91–95. Retrieved from Business Source Premier database.

Bednarz, A. (2005, July 5). The downside of offshoring. *Network World*, 33–35. Retrieved from Business Source Premier database.

Bednarzik, R. W. (2005, August). Restructuring information technology: is offshoring a concern? *Monthly Labor Review*, 128(8). Retrieved from ProQuest database.

Benamati, J. S., and Rajkumar, T. (2002, summer). The application development outsourcing decision: an application of the technology acceptance model. *Journal of Computer Information Systems*. Retrieved from Business Source Premier database.

Bowen, T., and La Monica, M (1998, August 17). IT gets picky with outsourcing. *Info World*, 43. Retrieved from Business Source Premier database.

Brandel, M. (2004, March 12). Offshoring grows up. *Computer World*, 25–27. Retrieved from Business Source Premier database.

Brooks, G. (2004, October 28). Reasons for outsourcing. *New Media Age*. Retrieved from Business Source Premier database.

Brooks, N. (2006, summer). Understanding IT outsourcing and its potential effects on IT workers and their environment. *Journal of Computer Information Systems*, 46–52. Retrieved from Business Source Premier database.

Buchanan, L. (2006, May). The thinking man's outsourcing. *Inc*, 28(5). Retrieved from Business Source Premier database.

Bureau of Labor Statistics (2008). Information technology–producing industries: extended mass layoff events and separations, private nonfarm sector, 2001–2008. Unpublished raw data. Retrieved September 15, 2008, from http://www.bls.gov/news.release/mslo.t06.html

Chang, J. C.-J., and King, W. R. (2005, summer). Measuring the performance of information systems: a functional scorecard. *Journal of Management Information Systems, 22*(1), 85–115. Retrieved from Business Source Premier database.

Cheon, M. J., Grover, V., and Teng, J. T. (1995). Theoretical perspectives on the outsourcing of information systems. *Journal of Information Technology,* 209–219. Retrieved from Business Source Premier database.

Clark, T. D., Jr., Zmud, R. W., and McCray, G. E. (1995). The outsourcing of information services: transforming the nature of business in the information industry. *Journal of Information Technology,* 221–237. Retrieved from Business Source Premier database.

Computer Economics (2010). Information Technoloyg employees can expect salary increase in 2010. Retrieved from http://www.cpst.org/hrdata/documents/pwm13s/C472S033.pdf.

Crockett, R. (2008). Plugging the gaps in U.S broadband. (2008, July). *Industry Week,* 60–61.

Datz, T. (2003, September 15). IT outsourcing: Merrill Lynch's billion dollar bet. *CIO.* Retrieved from Business Source Premier database.

Davis, B. (2004, March 29). Finding lessons of outsourcing in 4 historical tales. *The Wall Street Journal.* Retrieved from ProQuest database.

Davis, G. B., Ein-Dor, P., King, W. R., and Torkzadeh, R. (2006, November). IT offshoring: history, prospects and challenges [special section]. *Journal of the Association for Information Systems, 7*(11), 770–795. Retrieved from Business Source Premier database.

Dineley, D. (2001, February 12). Should outsourcers be part of your IT act? *InfoWorld.* Retrieved from Business Source Premier database.

Dix, J. (2006). Job skills for the new IT. *NetworkWorld*. Retrieved from Business Source Premier database.

Downing, C. E., Field, J. M., and Ritzman, L. P. (2003, winter). The value of outsourcing: a field study. *Information Systems Management*. Retrieved from Business Source Premier database.

Drezner, D. W. (2004, May/June). The outsourcing bogeyman. *Foreign Affairs*, 8(3), 22–24, 13. Retrieved from Business Source Premier database.

Dutta, A., and Roy, R. (2005, fall). Offshore outsourcing: a dynamic causal model of counteracting forces. *Journal of Management Information Systems*, 22(2), 15–35. Retrieved from Business Source Premier database.

Duvall, M. (2010). Global Outsourcing contracts set to soar. *CIO Zone*. Retrieved from http://www.ciozone.com/index.php/ Outsourcing/Global-Outsourcing-Contracts-Set-to-Soar. html

Earl, M. J. (1996, spring). The risks of outsourcing IT. *Sloan Management Review*, 37(3), 26. Retrieved from Business Source Premier database.

Eckle, J. (2005a, March 14). Most-sought IT skills. *Computer World*.

Eckle, J. (2005b, March 28). Offshoring's effects: mostly hitting low end. *Computer World*, 53. Retrieved from Business Source Premier database.

Eckle, J. (2008, July 14). Slowdown hits IT. *ComputerWorld*, 45.

Farrell, D. (2005, May). Offshoring: value creation through economic change. *Journal of Management Studies*, 42(3), 675–683. Retrieved from Business Source Premier database.

Feeny, D. F., and Willcocks, L. P. (1998, spring). Core IS capabilities for exploiting information technology. *Sloan Management Review*, 39(3), 9. Retrieved from Business Source Premier database.

Ferranti, M. (2003, November 15). Backlash of offshore outsourcing in the U.S. *CIO*. Retrieved from Business Source Premier database.

Fish, K. E., and Seydel, J. (2006, spring). Where IT outsourcing is and where it is going: a study across functions and department sizes. *Journal of Computer Information Systems*, 96–103. Retrieved from Business Source Premier database.

Fox, S. (2003). Economy. *Special Report on the U.S Economy.*

Fox, S. (2004, March 22). Offshore storm brews [letter to the editor]. *InfoWorld*, 11–12. Retrieved from Business Source Premier database.

Gibson, S. (2004, July 5). The offshoring tradition. *eWeek*, 31. Retrieved from Business Source Premier database.

Global Insight (2004). ITAA/Global insight study finds IT outsourcing results in net U.S. job growth. Retrieved September 15, 2008, from http://www.ihsglobalinsight.com/About/PressRelease/PressRelease855.htm

Gincel, R. (2005, July 4). Plotting your future in the global IT job market. *InfoWorld*, 29–33.

Gomberg, T. (2005, January). Globalization: a double edge? *Contract Management*, 45–46. Retrieved from Business Source Premier database.

Goodwin, B. (2006). Business focus needed as basic IT skills move offshore. *Computer Weekly*. Retrieved from Business Source Premier database.

Greene, W. (2006). Growth in services outsourcing to India: Propellant or drain on the U.S. economy? *US International Trade Commission*. Retrieved from Business Source Premier database.

Griffin, J. (2008). Bringing the business and IT strategies in line. *DM Review*, 33. Retrieved from Business Source Premier database.

Grupe, F. H. (1997, spring). Outsourcing the help desk function. *Information Systems Management*, 14(2). Retrieved from Business Source Premier database.

Handfield, W. (2006, October 18). Offshoring blamed for salary slump. *Computer Weekly*. Retrieved from Business Source Premier database.

Hauser, A. (2007, November). Offshoring industry faces new opponent: its clients. *Global Finance*, 6. Retrieved from Business Source Premier database.

Henderson, J. C. (1990, spring). Plugging into strategic partnerships: the critical IS connection. *Sloan Management Review*, 31(3), 7. Retrieved from Business Source Premier database.

Henley, J. (2006–2007, winter). Outsourcing the provision of software and IT-enabled services to India. *International Studies of Management & Organization*, 36(4), 111–131. Retrieved from Business Source Premier database.

Hickey, T. (2005, January). Outsourcing decisions: They're strategic. *Computer World*, 40. Retrieved from Business Source Premier database.

Hijzen, A., and Swaim, P. (2007, July). Does offshoring reduce industry employment? *National Institute Economic Review*, 201, 86–96. Retrieved from Business Source Premier database. doi:10.1177/0027950107083053.

Hirschheim, R., and Lacity, M. (2000, February). The myths and realities of information technology insourcing. *Communications of the ACM*, 43(2), 99–107. Retrieved from Business Source Premier database.

Hormozi, A., Hostetler, E., and Middleton, C. (2003, fall). Outsourcing information technology: assessing your options. *Sam Advanced Management Journal*, 18–23. Retrieved from Business Source Premier database.

Jurison, J. (1995). The role of risk and return in information technology outsourcing decisions. *Journal of Information Technology*, 10, 239–247. Retrieved from Business Source Premier database.

Kass, Elliot (2004). On Again. *Network World.* Retrieved from Business Source Premier database.

Kakumanu, P., and Portanova, A. (2006, September). Outsourcing: its benefits, drawbacks and other related issues. *The Journal of American Academy of Business*, Cambridge, 9(2), 2–6. Retrieved from Business Source Premier database.

Kaplan-Leiserson, E. (2004, October). Intelligence on offshoring. *T + D.* Retrieved from Business Source Premier database.

Kenney, B. (2008, January). Four keys to manufacturing IT's future. *IndustryWeek*. Retrieved from Business Source Premier database.

Klepper, R., and Jones, W. O. (1998). *Outsourcing Information Technology, Systems and Services* (S. Diehl, ed.). Prentice Hall PTR.

Lacity, M. C., and Hirschheim, R. (1993, fall). The information systems outsourcing bandwagon. *Sloan Management Review*, 35(1), 73. Retrieved from Business Source Premier database.

Lacity, M. C., and Willcocks, L. P. (1998, September). An empirical investigation of information technology sourcing practices: lessons from experience. *MIS Quarterly*, 363. Retrieved from Business Source Premier database.

Lee, J. -N., and Kim, Y. -G. (1999, spring). Effect of partnership quality on IS outsourcing success: conceptual framework and empirical validation. *Journal of Management Information Systems*, 15(4), 29–61. Retrieved from Business Source Premier database.

Lei, D. (1995, September/October). Strategic restructuring and outsourcing: the effect of mergers and acquisitions and LBOs on building firm skills and capabilities. *Journal of Management*. Retrieved from ProQuest database.

Levina, N., and Ross, J. W. (2003, September). The vendor's value proposition in IT outsourcing. *MIS Quarterly*, 27(3), 331–364. Retrieved from Business Source Premier database.

Loh, L., and Venkatraman, N. (1992). Information systems research. *The Institute of Management Sciences*, 334–358. Retrieved from Business Source Premier database.

Lohr, S. (2009). Troubles of Satyam could benefit rivals and 2 U.S. companies. *NY Times*. Retrieved from Business Source Premier database.

Lok, C. (2004, April). Where's my job? *Technology Review*, 74–75. Retrieved from Business Source Premier database.

Madden, S (2008). Make the most of your IT. Retrieved from Business Source Premier database.

Market Research Media (2010). U.S. Federal IT Market Forecast 2011-2015. Retrieved December 27, 2010, http://www.marketresearchmedia.com/2009/05/23/us-federal-it-spending-forecast-2010-2015/

Marks, S. (2003, December 22). Offshore outsourcing: business boom or bust? *Network World*, 20(51), 64. Retrieved from ProQuest database.

Mata, F. J., Fuerst, W. L., and Barney, J. B. (1995, December). Information technology and sustained competitive advantage: a resource-based analysis. *MIS Quarterly*. Retrieved from Business Source Premier database.

McDougall, P. (2005, August 8). India's next step. *Information Week*, 1051, 34. Retrieved from Business Source Premier database.

McGrath, M. (2007, January). Hot jobs and skills for 2007. *Certification Magazine*, 18–23.

Mears, J., and Bednarz, A. (2005, May). "Take it all" outsourcing on the wane. *Network World*, 22(21), 49.

Miller, M. (2004, May). Benefits of Offshore Outsourcing. *PC Magazine*, 9.

Mitchell, R. L. (2004). How IT has outsourced itself. *Computer World*. Retrieved from Business Source Premier database.

Mozumder, S. G. (2003, November 7). Outsourcing to bring long-term benefits to U.S. economy. *India Abroad*, 24. Retrieved from ProQuest database.

Mozumder, S. G. (2004, April 30). Anti-outsourcing laws may be unconstitutional. *India Abroad*, A25. Retrieved from ProQuest database.

Nayak, J. K., Sinha, G., and Guin, K. K. (2007, September). The determinants and impact of outsourcing on small and medium enterprises: an empirical study. *IMB Management Review*, 277–283. Retrieved from Business Source Premier database.

Neumann, W. L. (2008). *Understanding Research*. Boston: Pearson Education, Inc.

Nuwave Technologies (2010). Best Information technology job 2010. Unpublished raw data. Retrieved December 27, 2010, from http://www.nuwave-tech.com/it-project-blog/bid/45799/Best-Information-Technology-Jobs

Oberst, B. S., and Jones, R. C. (2006, June). Offshore outsourcing and the dawn of the post-colonial era of Western engineering education. *European Journal of Engineering Education*, 31(3), 303–310. Retrieved from Business Source Premier database.

Overby, S. (2003, September 1). The hidden costs of offshore outsourcing. *CIO*. Retrieved from Business Source Premier database.

Overby, S. (2004, January 15). Offshore outsourcing: how to safeguard your data in a dangerous world. *CIO*. Retrieved from Business Source Premier database.

Patterson, D. A. (2006, February). Offshoring: finally facts vs. folklore. *Communications of the ACM*, 49(2). Retrieved from Business Source Premier database.

PM Network (2007). All Ashore. *PM Network* 8. Retrieved from Business Source Premier database.

Pfannenstein, L. L., and Tsai, R. J. (2004, fall). Offshore outsourcing: current and future effects on American IT industry. *ISM Journal*. Retrieved from Business Source Premier database.

Purdum, T. (2007, September). Hidden costs of offshore outsourcing. *Industry Week*. Retrieved from Business Source Premier database.

Quinn, J. B. (1999, summer). Strategic outsourcing: leveraging knowledge capabilities. *Sloan Management Review*, 40(4), 9. Retrieved from Business Source Premier database.

Quinn, J. B., and Hilmer, F. G. (1994, summer). Strategic outsourcing. *Sloan Management Review*, 35(4), 43. Retrieved from Business Source Premier database.

Ramingwong, S., and Sajeev, A. (2007, August). Offshore outsourcing: the risk of keeping mum. *Communications of the ACM*, 50(8). Retrieved from Business Source Premier database.

RA Business (2010). New generation of MBA students. Retrieved from http://russianamericanbusiness.org/web_CURRENT/articles/590/1/New-generation-of-MBA-students

Robert Half Technology (2007). Salary Guide 2007. Retrieved from http://www.roberthalftechnology.com/SalaryCenter

Roberts, D. (2006). We must beware the offshoring skills wave. *Computer Weekly.* Retrieved from Business Source Premier database.

Ross, J. W., and Westerman, G. (2003, October). Architecting new outsourcing solutions: the promise of utility computing [editorial]. *Sloan School of Management*, 2–16. Retrieved from Business Source Premier database.

Rubin, H. (1997). Using metrics for outsourcing oversight. Retrieved from Business Source Premier database.

Sabherwal, R. (1999, February). The role of trust in outsourced IS development projects. *Communications of the ACM*, 42(2). Retrieved from Business Source Premier database.

Sen, F. (2006). Special issue on outsourcing. *Human Systems Management*, 89–90. Retrieved from Business Source Premier database.

Shearman, R. (2008, March). Keeping the business and IT relationship strong. *DM Review*, 41. Retrieved from Business Source Premier database.

Summerfield, B (2005, October). Who's afraid of Outsourcing? *Certification Magazine*, 38. Retrieved from Business Source Premier database.

Sobol, M. G., and Apte, U. (1995). Domestic and global outsourcing practices of America's most effective IS users. *Journal of Information Technology*, 269–280. Retrieved from Business Source Premier database.

Swartz, N. (2004, September/October). Offshoring privacy. *The Information Management Journal*, 24–26. Retrieved from Business Source Premier database.

Thibodeau, P. (2004a, September 27). GAO says offshoring could stymie tech job growth. *Computer World.* Retrieved from Business Source Premier database.

Thibodeau, P. (2004b, October 18). Offshoring fuels IT hiring boom in India. *Computer World*, 8.

Thibodeau, P. (2005, April 11). Costco aims to avoid offshore dependency. *Computer World*, 39(15), 7. Retrieved from Business Source Premier database.

Thibodeau, P. (2005a, May 2). Offshore tech support still stirs controversy. *Computer World*. Retrieved from Business Source Premier database.

Thibodeau, P. (2005b, July 4). Inaction on offshoring will hurt U.S. IT. *Computer World*. Retrieved from Business Source Premier database.

Thibodeau, P. (2006, February 6). India aims to tame soaring IT wages. *Computer World*, 15.

Thibodeau, P. (2010, December 6). Offshoring blamed in part of IT's jobless recovery. *Network World*.

TPI Index. (2009). Outsourcing contracts. Retrieved from http://www.allbusiness.com/company-activities-management/contracts-bids/12604706-1.html on September 15, 2008.

Tynan, D. (2006, September 18). Targeted training keeps workers sharp. *InfoWorld*. Retrieved from Business Source Premier database.

Venator, J. (2008, July). Entry-level certs fill skills gap. *Certification Magazine*, 22–23.

Venkatraman, N. (1997, spring). Beyond outsourcing: managing IT resources as a value center. *Sloan Management Review*, 38(3), 51. Retrieved from Business Source Premier database.

Venkatraman, N., and Henderson, J. C. (1998, fall). Real strategies for virtual organizing. *Sloan Management Review*, 40(1), 33. Retrieved from Business Source Premier database.

Vorholt, M. (2007, April). IT infrastructure managers: will the third wave energize companies? *Certification Magazine*. Retrieved from Business Source Premier database.

Warren, G. (2002). Outsourcing: A cost-effective alternative to the school help desk. Retrieved from Business Source Premier database.

Weber, R. (2004, June). Some implications of the Year-2000 era, Dot-com era, and offshoring for information systems pedagogy. *MIS Quarterly*, 28(2). Retrieved from ProQuest database.

Whitten, D. (2004–2005, winter). User information satisfaction scale reduction: application in an IT outsourcing environment. *Journal of Computer Information Systems*, 17–26. Retrieved from Business Source Premier database.

Willcocks, L. P., and Lacity, M. C. (1995). Information systems: outsourcing in theory and practice [editorial]. *Journal of Information Technology*, 10, 203–207. Retrieved from Business Source Premier database.

Worthen, B. (2003, September 1). The growing backlash of offshoring IT jobs. *CIO*. Retrieved from ProQuest database.

Xue, Y., Sankar, C. S., and Mbarika, V. W. (2004–2005, winter). Information technology outsourcing and virtual team. *Journal of Computer Information Systems*. Retrieved from Business Source Premier database.

About the Author

Mohammed Yusuf is an IT Account Manager in Manhattan, NY and a Professor in Patterson, NJ

Mohammed K. Yusuf has been a member of the Information Technology field for 10 years as an educator and a professional. Mohammed obtained his MBA from Nyack College in 2008, and has been teaching at local universities and college in the New Jersey area. An undergraduate in Management Information Systems has also contributed to his experience in the IT field and as an employee affected by the effects of offshore outsourcing, he spent two years researching and developing this book for his graduating thesis from Nyack College. He has always dreamed of publishing information from my research to educate others in the education space, and defunct the myths around offshore outsourcing. He came into this as a believer in those myths, however my years of research has lead me to keep an open mind and view offshore outsourcing for its benefits through globalization. He currently reside in New Jersey, with my wife and son. He enjoys photography, technology and most especially soccer. This is his first novel.